FOREWORD

The Joint Staff envisions the future operating environment as one characterized by uncertainty, complexity, rapid change and persistent conflict. The continued spread of low-cost, high-technology information systems could soon present the U.S. with an array of technological peers in a relatively short period of time. Global advances in the areas of long-range precision weapons, unmanned systems, and cyberspace weapons will significantly complicate U.S. planning efforts and could constrict U.S. military freedom of action across a number of warfighting scenarios.

The essential access challenge for future U.S. military forces will be our ability to project force into a heavily-contested operational area and to sustain the force in the face of increasingly capable enemies equipped with sophisticated anti-access and area-denial defenses. Over the next fifteen years, the impact of these emerging technologies will lead to an increasingly complex and more lethal maritime operating environment. Of concern, some adversaries are already developing the military means to impede or prevent U.S. naval forces from responding to crises overseas. In particular, growing threats within the cyberspace domain and across the electromagnetic spectrum could soon challenge long-held U.S. Navy advantages within selected areas of the maritime battlespace.

Projected operational and informational environments are driving the need for significant changes and improvements in how the Navy will use and protect its current, planned and forecasted information-based capabilities in conflict. This U.S. Navy Information Dominance Roadmap, 2013–2028 was developed to highlight emerging challenges across the range of military environments, and to describe Navy's advanced information-based capabilities required in the areas of Assured Command and Control, Battlespace Awareness and Integrated Fires. This roadmap is intended to help synchronize and integrate Navy's diverse information-related programs, systems, functions and initiatives to maintain decision superiority and meet future combat objectives in high-threat environments. Achieving these advanced capabilities will require leveraging Navy's intellectual, technological, human and financial resources across the Fleet, Systems Commands and OPNAV Staff.

Executive Summary

The Navy is pursuing improved information-based capabilities that will enable it to prevail in the higher-threat, information-intensive combat environments of the 21st Century. This document outlines challenges anticipated over the next 15 years in the operating and information environments, and highlights long-term opportunities for fully integrating Navy's information-related activities, resources, processes and capabilities to optimize warfighting effects and maintain decision superiority across the spectrum of warfare. The Navy's plans for achieving Information Dominance center on: 1) assuring command and control (C2) for our deployed forces regardless of the threat environment; 2) enhancing battlespace awareness to shorten the decision cycle inside that of the adversary and to better understand the maritime operating environment; and, 3) fully integrating traditional kinetic and emerging non-kinetic fires to expand warfighting options to both Navy and Joint commanders. To accomplish these plans, today's current information-based capabilities involving Assured C2, Battlespace Awareness, and Integrated Fires will require continual changes and improvements in a number of diverse areas.

Assured C2 will require a more robust, protected, resilient and reliable information infrastructure that undergirds the Navy's overall information environment and allow uninterrupted worldwide communication between deployed units and forces ashore. Navy's information infrastructure must be able to maintain essential network and data link services across secured segments of the electromagnetic spectrum in order to transport, share, store, protect and disseminate critical combat information.

Battlespace Awareness will require enhanced information content, advanced means to rapidly sense, collect, process, analyze, evaluate and exploit intelligence regarding our adversaries and the operating environment. Our information content will serve as the basis from which nearly all decisions will be made, enabling our forces to more effectively maneuver and coordinate actions that target and engage enemy forces.

Integrated Fires will require new capabilities to fully employ integrated information in warfare by expanding the use of advanced electronic warfare and offensive cyber effects to complement existing and planned air, surface and subsurface kinetic weapons within the battlespace. Future information effects will be designed to impact and change adversary behavior, or when necessary, to control, manipulate, deny, degrade or destroy his warfighting capabilities.

TABLE OF CONTENTS

Navy Information Dominance is defined as the operational advantage gained from fully integrating Navy's information capabilities, systems and resources to optimize decision making and maximize warfighting effects in the complex maritime environment of the 21st Century.

INTRODUCTION

Purpose: This document summarizes the operating and information environments expected during the 2013–2028 timeframe and depicts Navy's required future Information Dominance capabilities. This is intended to help synchronize and integrate Navy's diverse information-related programs, systems, functions and initiatives to meet future warfighting objectives. This document expands on the initial goals and objectives contained within the Navy Strategy for Achieving Information Dominance, 2013–2017; however, it focuses more on longer-term Navy planning and resourcing decisions for developing the information-based warfighting capabilities and capacities the Navy will require to maneuver freely in future high-threat, information-intensive environments at sea.

Scope: This roadmap addresses information-related capabilities and activities under Navy's control that involve decision making and warfighting within the battlespace (i.e., understanding the enemy, networking the force, shortening combat kill chains, expanding warfighting options, etc.). Navy's information-based systems that pertain to personnel management, budgeting, contracting, logistics, etc. are not addressed in this document.

Information Dominance Defined: Navy Information Dominance is defined as the operational advantage gained from fully integrating Navy's information capabilities, systems and resources to optimize decision making and maximize warfighting effects in the complex maritime environment of the 21st Century. The development of a Navy-wide Information Dominance capability is being driven by trends within the worldwide information and operating environments, which are predicted to stress U.S. Navy freedom of movement and capabilities in future conflict.

Current/Near-Term Operational Environment (2013–2019):

Today's steady-state operational environment can be characterized as one of increasing competition and rapidly changing technological advances. U.S. military operations take place amidst a backdrop of rising military powers, a diffusion of military capability flowing to non-state actors, and greater access by individuals and small groups to lethal weapons, sensors and other technologies. In this environment, the U.S. Navy freely operates as the dominant maritime power. Moreover, the U.S. military enjoys superiority on the land, air and sea, as well as in the space and cyberspace domains, and this U.S. asymmetric advantage is due in large part to our assured C2 capabilities. However, rapid advances being made by other nations in science and technology, expanded adoption of irregular warfare tactics by both state and non-state actors, proliferation of long-range precision weapons, and the growing use of cyber attacks are all increasingly constricting U.S. military freedom of action.

Adversary Observations: Our adversaries recognize that the U.S. military possesses significant advances in many technologies, and are actively pursuing counter-measures to mitigate these advantages. Observations of U.S. military operations in Iraq and Afghanistan since 9/11 have led some adversaries to conclude that:

- Anti-access, access limitation, and tactical shielding can be effective means in slowing U.S. response to crises and controlling our ability to react in theater;
- U.S. high tech weapons and communications systems can be countered with low-technology responses;
- Small numbers of sophisticated weapon systems can have a dramatic effect on the operational environment.

Current/Near-Term Information Environment (2013–2019):

Nations are facing rapidly changing and increasingly complex information and cyberspace environment. Emerging information technology (IT) systems are often being developed and implemented faster by the civilian sector than by worldwide governments. In this setting, the U.S. Navy continues to operate from the information "high ground," employing superior intelligence and network technologies faster than our adversaries. However, the Navy's relative advantages are eroding steadily as some adversaries are now actively exploiting modern information-based capabilities and technologies for their own use. In some areas, our adversaries are beginning to develop information-based offensive and defensive capabilities that rival our own. In other areas, many are simply using the internet and the commercial global information grid as their own C4ISR system for networking their low-technology military forces.

Future Operational Environment (2020–2028):

According to the National Intelligence Council's *Global Trends 2025: A Transformed World* and *Global Trends 2030: Alternative Worlds*, the future international system will be almost unrecognizable from previous decades owing to the rise of emerging powers, an increasingly globalized economy, an unprecedented transfer of relative wealth and economic power flowing to Asian states, and the growing influence of several non-state actors. The U.S. will remain the single most powerful country in the late 2020's but will be less dominant on the world stage, and will see its relative strength—even in the military realm—decline. The ongoing shift in relative wealth and economic power flowing from West to East is expected to continue. The world will move towards an even more globalized, multi-polar international system where long-standing gaps in power between developed and developing countries will narrow. Of primary concern to U.S. planners, such multi-polar international systems have historically been more unstable than bipolar or uni-polar ones, suggesting the next twenty years of transition to a new world order will be fraught with instability and risks. Within this future international system, the U.S. Navy will continue to play a significant role in global maritime security.

Operational Trends: The Joint Staff Capstone Concept for Joint Operation (CCJO) envisions the future operating environment as one characterized by uncertainty, complexity, rapid change, and persistent conflict. According to the Joint Staff, three emerging trends will increasingly complicate the U.S. military's ability to access key operational areas in times of conflict. Listed below, these could impact or jeopardize Navy's success in high-threat operating environments:

- The dramatic improvement and proliferation of long-range weapons and other technologies capable of denying access or freedom of action within an operational area.
- The emergence of space and cyberspace as contested warfighting domains.
- A defense posture where fewer U.S. basing rights overseas will be available for ground forces, while the Navy increases its overseas footprint.

Anti-Access/Area Denial Capabilities: The Joint Staff Joint Operational Access Concept (JOAC) indicates that the essential access challenge for U.S. military forces will be our ability to project force into a contested operational area, and to sustain it in the face of armed opposition by increasingly capable enemies equipped with sophisticated anti-access and area-denial (A2/AD) defenses. Advanced A2/AD capabilities and weapons are increasingly becoming available to both less-developed states and non-state actors. By 2020, numerous nations and non-state actors will have the military means to selectively deny access to key maritime chokepoints and other strategic areas. From a maritime standpoint:

- Adversary Strategies for countering U.S. Navy strengths could include conducting simultaneous operations to temporarily overwhelm Fleet defenses, reducing time delays in coordinating their attacks, and exercising better control of their own strategic information environment.
- Adversary Operations for countering U.S. Navy strengths could include avoiding our strengths, exploiting the physical environment as well as our overseas operational constraints, seeking "game-changing" technologies for use in the battlespace, and disrupting U.S. naval operations by using all available elements of their informational and technological capabilities.
- Adversary Tactics could include exploiting asymmetric advantages to gain synergy over our conventional modes of operation, employing the full range of emerging technologies in warfare to include cyber and advanced electronic warfare, and utilizing information warfare as a key weapon system.

Maritime Operating Environment: The maritime operational environment will become increasingly complex and more lethal. Longer-range and more sophisticated manned and unmanned sensors and anti-ship weapons are expected to proliferate, further complicating U.S. Navy planning efforts. Some adversaries are already working on capabilities to slow or prevent U.S. forces from responding to future conflict or aiding our overseas allies and partners. A few adversaries are mounting strategies to prevent U.S. forces from even entering a theater, or operating effectively once there. Forecasts from the Office of Naval Intelligence and the Joint Staff's *JOAC* conclude that:

- Both state and non-state actors will continue to pursue A2/AD strategies involving advanced mines, submarines, anti-ship cruise and ballistic missiles, anti-satellite weapons, cyber attacks, and communications jamming.
- The current and projected proliferation of technology and high-tech weapons worldwide may allow some enemies to achieve limited or unexpected parity (and possible superiority) in selected military technologies.
- Unforeseen technological surprises can also be expected and will likely become more common as technological advances proliferate, further impacting future U.S. military capabilities and options.

Future Information Environment (2020–2028):

Rising powers in Asia are poised to have tremendous impact on the world over the next 15–20 years. Some will be in optimal position to develop emerging breakthrough technologies to include robotics, nanotechnology, and the next generation of the Internet. The proliferation of advanced information technology could hinder U.S. efforts to maintain future access to the cyberspace domain. The rapid pace of scientific breakthroughs in information technology will continue to accelerate. Joint Forces Command's *Joint Operating Environment (JOE)* postulates that such advances will change the very character of war. Highlights include:

Information Technologies: Information and communication technology, advanced electronics, bioengineering and nanotechnology will all have profound effects on military operations in the coming years. Some scientific advances may well redefine many dimensions of civilian society. Developments in quantum computing and nanotechnology could lead to a fighting force further enhanced by improved robotics and remotely-guided autonomous and miniaturized weapons, all supported by advanced communications systems that will become more self-organizing and distributed. Forecasts from the *Army's Operational Environment (OE) 2009–2025* indicate that:

- By 2015, computer chips will have evolved from silicon transistors to nanomaterials.
- Computing power is expected to grow exponentially in the next 10–20 years, with computing speeds over 1,000 times faster than today's supercomputer.
- New data storage techniques such as "nanotechnology-enabled memory" will vastly increase the capacity to store and transmit data.
- In terms of bandwidth (ashore), supply will exceed demand.
- By 2025, given other advances in science and technology, quantum computing will be possible.

Dual-use Information Technologies: The impact of rapidly expanding global and regional information architectures, systems, and organizations, both private and public, cannot be overstated. Relatively low-cost, dual-use civilian technologies involving high-resolution imagery, information transfer and display, and global positioning systems are widely available on the commercial market and will continue to proliferate. Such dual-use technology transfers have already proven invaluable to those actors seeking to upgrade their weapons systems or facilitate research and development efforts. This spread of low-cost, high-technology commercial systems could present the U.S. with an array of future technological peers in a relatively short timeframe. Other forecasts from *OE 2009–2025* include:

- With advanced information technology, weapons proliferation (to include conventional, cyber and weapons of mass destruction) will spread and could improve the disruptive and destructive capabilities of a number of state and non-state actors.
- Such proliferation may allow adversaries, be they non-state or rising state actors, to achieve limited parity or even superiority in selected niche technology areas.
- Ubiquitous sensors of various types and sizes could soon saturate the operational environment, and these capabilities will eventually be available on the world market for a price, giving any threat resourced to purchase them the means to enhance their strike capabilities.

Information Vulnerabilities: Future technological advances could also have adverse effects on advanced military forces, like the U.S. Navy, that are heavily reliant on technology to operate. The vulnerabilities of information technology will constantly grow through the continued global development and transfer of ever evolving and accelerating cycle of IT security measures and countermeasures. Adversaries will continue their attempts to counter the long-held U.S. advantage in communications, surveillance and long-range precision fires. Left unchecked, the growing vulnerabilities of our information systems will have consequences and will require deliberate changes in how the Navy employs and protects its information. *OE 2009–2025* highlights regarding potential adversaries responses to U.S. information-based strengths include:

- Threat actors understand U.S. reliance on communications, ISR, and visualization technologies, and perceive them as vulnerable to disruption and exploitation.
- One of the most important areas in which threat actors will seek to operate is in the acquisition and employment of sophisticated EW systems.
- An adversary able to cause significant disruption or degradation to our global positioning satellites could dramatically reduce our precision strike advantage.

U.S. 5TH FLEET AREA OF RESPONSIBILITY (Nov. 11, 2012) USS Farragut (DDG 99) underway.

- An adversary with the ability to intercept and manipulate satellite communications could disrupt or block information required for decision making in combat.
- Attacks could well include cyber and the employment of electromagnetic pulse (EMP) devices to disrupt internal kill chains, impact supply chains and infiltrate network resources.

Space and Cyberspace Vulnerabilities: As a forward deployed force, the Fleet is highly dependent upon space-based systems, cyberspace and the EM spectrum. Modern wars involve the exploitation of cyberspace, and future wars between advanced nations may include offensive space operations. While the U.S. has enjoyed uncontested superiority in space for several decades, ever cheaper access to space and the emergence of anti-satellite and counter-space weapons have begun to level the playing field. As more nations and non-state actors develop counter-space capabilities, threats to U.S. space systems and challenges to the stability and security of the overall space environment will increase. Forecasts from the *JOE - 2010* include:

- Potential adversaries will exploit perceived U.S. space and cyberspace vulnerabilities, which could impact U.S. information-handling capabilities and wartime readiness.
- Cyberspace threats already pose a critical national and economic security concern to the U.S. due to our dependence on information in nearly all aspects of society.
- Data collection, processing, storage, and transmission capabilities are increasing exponentially.
- Mobile, wireless, and cloud computing will bring the full power of the globally-connected Internet to myriad personal devices and critical cyberspace infrastructures.

Information Dominance Capabilities for the 2013 to 2028 Timeframe:

The projected operational and informational environments outlined above are driving the need for significant changes and improvements in how the Navy will use and protect its current, planned and forecasted information-based capabilities. Achieving Information Dominance under the three areas of **Assured C2, Battlespace Awareness** and **Integrated Fires** will require significant changes and improvements in the Navy's approach to managing its information infrastructure, content and effects, respectively. Navy's approach for advancing its future information-based capabilities are further outlined and described in the three chapters that follow.

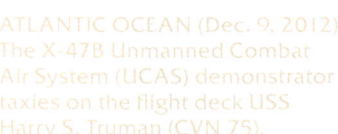
ATLANTIC OCEAN (Dec. 9, 2012) The X-47B Unmanned Combat Air System (UCAS) demonstrator taxies on the flight deck USS Harry S. Truman (CVN 75).

CHAPTER

ASSURED COMMAND AND CONTROL (C2)

NETWORK THE FORCE IN ANY ENVIRONMENT

Overview: Naval commanders must be able to exercise authority over assigned and attached Navy and Joint forces to ensure success across the range of military operations and warfighting scenarios. Continued improvement in the speed and sophistication of modern information technology is greatly increasing the capability of naval forces to coordinate action across disparate sea, land, air, space, and cyberspace systems. For the U.S. Navy, such advancements have already increased decision space for commanders afloat, providing tremendous improvements in offensive capability. Adversaries, however, are actively working to degrade or negate the Navy's operational advantage by developing the means to disrupt C2 systems and impair our ability to direct and coordinate actions. In the increasingly hostile operational environments expected in the future, the Navy must be prepared to defend its network and navigation systems and control key segments of the electromagnetic (EM) spectrum. The concept of Assured C2 seeks to maintain the Navy's ability to exercise C2 in the presence of a protracted "information blockade" employed by adversaries, especially under heavily contested or denied operational conditions. The range of C2 environments the Navy expects to face in the future include:

- **Permissive C2.** Most Joint missions are conducted under this environment. In such scenarios, the Navy's communication and networking infrastructure is sufficient to network the force and enable freedom of action. Bandwidth and channel capacity, however, may be strained in large operations involving many users. *Threats to the Navy's ability to manage its information and networks in this environment involve the spread of computer viruses and worms, hacking, unauthorized downloading of large data files, and the interception of unencrypted own force radio frequency (RF) signals.*

- **Contested C2.** An escalation of hostilities could lead to an environment where Navy forces face growing threats to their networking, satellite communications (SATCOM), and Global Positioning System (GPS) capabilities. In spite of such threats, naval forces would maintain at least one communications path for operational purposes. *Threats to the Navy's capabilities in this environment involve intermittent degradations of SATCOM and RF links, and momentary interruptions of GPS signals (but with no significant degradation to precision navigation).*
- **Highly Contested/Denied C2.** Further escalation could lead to a highly contested or even denied C2 environment where forces face a near total loss of their commercial and military-specific networking capabilities due to adversary action. Forces will be challenged to provide even one communication path for most information requirements. *Threats to the Navy's capabilities in this environment potentially involve a total loss of existing SATCOM and RF links, as well as a loss of GPS position navigation signals.*

Functional Description:

Under the threat environments listed above, the three primary warfighting functions associated with Navy Assured C2 that must be accomplished are:

Command Forces in Any Environment. Navy commanders must be able to plan, direct, lead and coordinate forces and operations under all conditions, regardless of threat. Such a capability requires a robust baseline information grid that continually connects the commander to the force, connects units to higher echelon C2 and combat support, defends the network, provides assured access to the EM spectrum, and provides accurate positioning, navigation and timing (PNT) services. Forecasted changes in the evolving networking environment require the Navy to: 1) maneuver freely in the congested EM spectrum to assure continued C2; 2) establish a dynamic and flexible information grid; 3) deliver mission-relevant data for shared awareness; and, 4) assure timing for the information infrastructure as well as tactical coordination.

Coordinate Fires in All Domains. Naval commanders must maintain the means to plan, maneuver the force, and coordinate fires for kinetic and non-kinetic effects. By seizing and retaining the initiative, we force our adversaries to react to our actions and conform to our plans. Such a capability requires: 1) a data-centric access to essential combat information; 2) processing services and interfaces to support coordinated planning, execution decision making and dynamic battle management; 3) fast, reliable and secure tactical networks that link platforms, sensors and weapons; and, 4) assured PNT services. This will require the Navy to enhance operational and tactical decision support; enhance data link networks supporting tactical execution; and assure navigation for tactical platforms and weapons.

Assess Fires and Own Force Status. Naval commanders must be able to report and receive timely assessments on the mission status, strike results, enemy forces, neutrals/non-combatants, friendly forces, terrain and weather under any conditions. This will require the Navy to maintain real-time assessments of operational readiness, and adopt technologies to improve dissemination of combat data.

Current State/ Planned Milestones (2013-2019):

Over the next six years, a number of changes are programmed to improve the Navy's ability to assure C2 in current and near-term warfighting scenarios. Highlights over the 2013–2019 timeframe regarding Navy AC2 capabilities are characterized as follows:

- Navy investments in satellite-based communications will provide viable, high-bandwidth, overthe-horizon transport options for connecting forward deployed naval units at the tactical edge with C2 Forward, C2 Rear and other supporting C2 nodes ashore in a permissive environment.
- A warfighting "thin-line" will be delivered to sustain critical communication paths in contested environments using protected SATCOM with split-Internet Protocol (IP) broadcast and tactical high frequency (HF) data networks, and by manually shutting down non-critical systems. In addition, low bandwidth line of sight (LOS) communications will be relayed through surface, subsurface, or air layers to support transport requirement.
- The Navy expects that the Consolidated Afloat Networks and Enterprise Services (CANES) program will begin to deliver the next generation Navy tactical network with a Common Computing Environment and Afloat Core Services (ACS) enabling information sharing and a common understanding of the battlespace.
- The Next Generation Enterprise Networks (NGEN) is beginning to provide secure, net centric data and services for the Navy and Marine Corps. NGEN will be the foundation for DON network consolidation, and will be interoperable with, and able to leverage, DoD Joint Information Environment (JlE) services.
- The Navy expects that initial deployments of improved Tactical Data Link (TDL) networks will begin to address current capability and capacity limitations.
- The Navy will be able to maintain accurate PNT capabilities from satellite-based GPS signals is a range of operational environments.
- The Navy will begin to install limited EM spectrum controls to ensure the transport infrastructure delivers reliable high-bandwidth connections to naval platforms, enabling a number of warfighting applications. Fleet commanders at the Maritime Operations Center (MOC) level can establish and share a basic maritime current operational picture (COP) that is periodically updated, and employs limited COP tools to assist planning and execution decision making. Several data interfaces will exist between Navy combat systems and C2 networks.

Advanced Capabilities (2020-2028):

Advanced improvements and new capabilities will be required to maintain the Navy's future ability to assure C2 in degraded and/or denied C2 environments. Required Navy AC2 capabilities in the 2020–2028 timeframe are characterized under the major functional areas outlined below.

Command Forces in Any Threat Environment:

1) Assured EM Spectrum Access: To command forces in high threat environments requires common systems data schema to enable monitoring and manipulation of each system's EM spectrum inputs and outputs. This involves developing an EM spectrum awareness and control capability that enables

ATLANTIC OCEAN (June 30, 2011) USS California (SSN 781) underway during sea trials.

individual platforms to manipulate and adjust emissions to maximize operational capability. This would reduce vulnerability to detection or jamming, and ensure warfighter communication paths are always available. This will require innovative techniques and adaptive RF solutions that enable sharing of the entire EM Spectrum (including federal and non-federal frequency bands), and leverage gray and white space technology to give the Navy the EM maneuver space it requires to execute war plans under any threat condition. Such advanced capabilities involve:

- The ability to field increased numbers of LOS communication systems for tactical operations with a common set of shipboard RF apertures and components for communications, EW and navigation radars;
- The ability to monitor selected combat systems for their operational status and adjust them via automated operations;
- A COP of the EM spectrum that is linked to electronic navigation charts and displays operational restrictions to enable dynamic spectrum control;
- A fully functioning Real-Time Spectrum Operations (RTSO) capability that enables dynamic monitoring and control of EM spectrum emissions from every strike group and platform, and reduces spectrum operational response time from minutes to seconds.

2) A Dynamic Flexible Grid: A dynamic flexible information grid will be required in future environments to ensure that every node (platform, sensor, weapon system) can connect to and extend the grid, either through SATCOM, an aerial layer with Mobile Global Information Grid (GIG) Entry Points (MGEP), or by point-to-point tactical grid connections. Such a capability will provide the warfighter an agile transport environment by reducing reliance on the space layer for wideband connectivity, and by providing a "C2 thin-line" to enable communications in denied environments. A dynamic grid should be able to sense and react to changes in the operational environment, and re-establish critical communications between any available mobile assets across the enterprise. Such advanced capabilities involve:

- An agile routing capability that dynamically routes IP traffic over multiple paths between tactical edge platforms;
- Enterprise management of grid capacity based on changing mission needs;
- Automated information sharing interfaces across combat, C2 and ISR systems.

- Reduced Navy reliance on fixed teleport sites (i.e., establishing MGEP's);
- A network control capability at the tactical edge that reduces reliance on shore based network control and allows the network to automatically resynchronize with the GIG and shore-based services when connections are re-established);
- An EMS operations capability that provides operational status and readiness of all platforms within the Strike Group, and allows a commander to centrally control strike group emissions, depending on the operational situation.

3) Mission-Relevant Data for Shared Awareness: A more advanced data centric architecture will be required in future environments to allow rapid extraction of operational, intelligence, meteorological/oceanographic, EM spectrum and network health data whenever needed. Such a capability will require assured EM spectrum access, the ability to protect both CS and C4I networks, and improved understanding of environmental characteristics. Such advanced capabilities involve:

- A capability for key systems to automatically report their operational status via their inherent architectures and new data schemas that enable systems to push-pull required information within a common environment;
- A force discovery service that registers newly arriving network participants and their operational requirements, and displays their status on a Navy Tactical Display System (NTDS) type display for Tactical Action Officers (TAOs) and fleet commanders.

4) Assured Timing Services: Alternative methods of ensuring synchronized time and frequency in GPS degraded or denied environments are needed to operate effectively in highly contested or denied C2 environments. Such advanced capabilities involve:

- Enhanced redundancy to existing space-based communications, making them highly resistant to ground-based jamming threat (i.e., two-way satellite time transfer);
- A low frequency navigation system (i.e., LORAN) for time and frequency synchronization.

Coordinate Fires in all Domains:

1) Operational and Tactical Decision Support: Commanders at all levels will require the ability to rapidly discover and access key relevant intelligence and operational data to maintain decision superiority in contested or denied C2 environments. Such advanced capabilities involve:

- Advanced decision support aids that can provide real-time sharing and collaboration of authoritative intelligence and operational data across all levels of wars;
- Tailored applications for the fleet commander that would enable visibility of key information and data attributes to enable prioritization for dissemination, and unit-level tactical applications that would enhance dynamic battle management decision making and execution;
- An enhanced information layer that enables all-source data to be ingested, tagged, stored, organized and shared for operational analysis, and accessible and discoverable between tactical, operational and national data sets;

- Improved computational capacity, artificial intelligence and tools to enhance situational awareness and enable rapid modeling of alternate operational and tactical courses of action.

2) Enhanced Data Link Networks: The Navy requires improved networked track data speed, capacity, reach and reliability. Disparate Navy systems need to be network-enabled so that platforms, sensors, weapons and systems can seamlessly exchange combat data and information via machine-to-machine interfaces. Such advanced capabilities involve:

- Dynamic mesh networks that provide speed and capacity not available with the current Link-16 system;
- Advanced tactical targeting network technology that integrate the E-2D with other airborne platforms and surface/subsurface sensors to increase the capability to track and engage targets in dense threat environments;
- Advanced TDL waveforms that will improve track capacity and enhance other performance characteristics;
- New tactical data links extending to the undersea environment;
- New low probability of intercept (LPI) and low probability of detection (LPD) acoustic links that enable low data rate, long-range communication between undersea platforms in clandestine operations;
- Improved sensitivity of gyroscopes and inertial systems to accurately capture attitude and orientation in three dimensions and enable improved stellar/celestial navigation capabilities.

3) Assured Positioning, Navigation and Timing (PNT): To operate effectively in future maritime environments, the Navy must also maintain assured synchronized timing and frequency capabilities, which will require more precise timekeeping, better frequency stability, a smaller size, and lower power demands. Such advanced capabilities involve:

- New protected GPS signals and receivers and advanced GPS anti-jam antennas that can ensure continued PNT availability;
- Alternative navigation data, such as stellar navigation or improved inertial information, to augment GPS navigation capabilities;

U.S. 5TH FLEET AREA OF RESPONSIBILITY (Jan. 2, 2013) USS John C. Stennis (CVN 74), USNS Bridge (T-AOE 10), and USS Mobile Bay (CG 53) (foreground) conducting theater security cooperation efforts.

- More robust, redundant and assured three-dimensional PNT capabilities that enable hardened and protected PNT sources;
- A modified COP that displays positioning and navigation data and spectrum operational restrictions (overlays) which instruct or cue other systems to change frequency assignments and channels when needed, based on their location;
- An acoustic data transfer capability for existing PNT data and improved pressure gradient sensors that provides more accuracy in underwater position and navigation;
- Improved accuracy and synchronized atomic referenced oscillators that maintain required PNT accuracies over longer durations;
- Spectrum-borne, non-organic ubiquitous timing reference signals.

Assess Fires and Own Force Status:

1) Timely Assessments of Operational Readiness: In future highly contested or denied C2 environments, commanders will require an improved capability to continuously assess the result of their blue fires and own force status. This can be enabled by aligning, integrating, and transforming the Navy's current networking capabilities and strategies in such a way as to "adapt to, or adopt" the evolving JIE network. To support this data-centric goal, the Navy must align with DoD's major governance areas (requirements, budget, acquisition, and operations oversight) and coordinate other Service and agency IT efforts into the development of an optimized DoD plan. This will enable shared situational awareness and provide advanced decision support tools (i.e., red course of action, EM interactions, effects prediction, etc.). Such advanced capabilities involve:

- Combat systems able to automatically report their operational readiness and pass information to update an operationally-relevant COP of friendly and other service forces;
- Improved tailored displays with automated updates on all aspects of force readiness for multiple missions;
- New capabilities to leverage emerging ISR, BA and IF capabilities to provide visual operational capabilities across the battlefield;
- Self-healing and aware networks systems able to adjust and reconfigure connections as needed;
- An identity and access management capability allowing a global network access to all users.

2) Adoption of Smart Sensor Data Dissemination: Navy will require a smart sensor grid enabling data to be disseminated rapidly across the battlespace in future contested or denied C2 environments. Such advanced capabilities involve:

- A network infrastructure that supports employment of a smart sensor strategy, sensors that process data locally and disseminate required data smartly, and netted sensors that sense, store, assimilate process, communicate and transport essential data;
- Netted sensors able to sense, store, assimilate, process, communicate and transport essential data in congested EM environments;
- Enhanced means to operate in a cloud-enabled environment using devices that allow secure and cost effective operations at the point of need utilizing all available transport assets (sea, undersea, air and land).

PACIFIC OCEAN (Jun. 30, 2012) 55 feet remain visible after the crew of the Floating Instrument Platform, or FLIP, partially flood the ballast tanks causing the vessel to turn stern first into the ocean. The 355-foot research vessel, owned by the Office of Naval Research and operated by the Marine Physical Laboratory at Scripps Institution of Oceanography at University of California, conducts investigations in a number of fields, including acoustics, oceanography, meteorology and marine mammal observation.

CHAPTER 2

Battlespace Awareness

KNOW THE ENEMY, KNOW THE ENVIRONMENT

Overview: Battlespace Awareness (BA) is the ability to understand the disposition and intentions of potential adversaries as well as the characteristics and conditions of the operational environment. This knowledge impacts Navy and Joint planning, operations and decision making at the strategic, operational, or tactical level. The Navy's operational environment spans all domains (maritime, air, land, space and cyberspace) and all frequencies across the EM spectrum. Navy BA relies upon Navy's Assured C2 capabilities, enables Navy Integrated Fires, and provides naval commanders with the level of decision superiority required to execute the broad array of Navy missions. Effective BA within the Navy must leverage all available sources of information, and requires a profound knowledge of the following maritime-related areas: 1) potential adversary locations, activities, intent and capabilities, including traditional, asymmetric, cyber, and emerging methods of warfare; 2) Joint, coalition, neutral party and own force capacity, capability and status; and, 3) the physical and virtual environments and their potential impact on mission execution. The major elements of BA and he corresponding Navy concerns in each of these areas are highlighted below:

- **Tasking, Planning, and Direction:** The ability to synchronize and integrate the activities and resources of collection, processing, analysis, and dissemination to meet information requirements. Navy's capabilities in this area are constrained by challenges relating to the tasking of non-organic sensors and assets, coordinating with all external stakeholders, and measuring the mission impact of various BA-related activities.
- **Collection:** The ability to gather and obtain required data to satisfy information needs. *Navy's capabilities in this area are constrained by difficulties in*

integrating and coordinating organic and non-organic sensors and platforms, challenges with monitoring non-cooperating platforms across multiple domains, and gaining sensor access to denied areas.

- **Data and Information Sharing:** The ability to share data and information at all classification levels with national and international partners. *Navy's capabilities in this area may be overwhelmed by increasing volumes of sensor data streaming from Navy and Joint assets, the growing number of stakeholders with whom Navy shares information, immature capabilities in managing large data stores, and the need for more closely-coordinated cross-domain information systems with our international partners.*

- **Processing and Fusion:** The ability to transform collected information into forms suitable for further analysis and/or action by man or machine. *Navy's capabilities in this area need to be enhanced to deal with increasing volumes of data.*

- **Analysis, Prediction, and Production:** The ability to integrate, evaluate and interpret knowledge and information from available sources to develop more predictive intelligence and forecast the future states of the physical and virtual environments to enable situational awareness and provide actionable information. *Navy's efforts in this area are hampered by unclear roles and responsibilities among various Navy entities engaged in these activities, and the tendency to analyze and maintain various types of intelligence data within their separate disciplines.*

- **Information Dissemination and Management:** The ability to present and make available intelligence, information, and environmental content that facilitates understanding of the operating environment by military and national decision-makers. *Navy's capabilities in this area do not allow for adequate information sharing and visualization, especially across multiple domains.*

Functional Descriptions:

The three primary warfighting functions associated with Navy BA are briefly outlined below:

Fuse Essential Combat Information: Navy commanders require immediate and continual access to essential combat information in order to maneuver the force and execute the full range of missions under a broad spectrum of operating conditions. To sustain the flow of combat information within A2/AD environments, the Navy will be required to streamline tasking, planning and direction; advance sensor development across all domains; and, automate processing, fusion and product delivery.

Understand the Operating Environment: Navy commanders base their operational decisions on a continually evolving understanding of the operating environment which relies on the accuracy and completeness of information available at any given point in time. To optimize the utility and value of essential information, the Navy will be required to develop a shared, relevant real-time COP; comprehend and predict the physical and virtual environments; and understand the capabilities and intentions of allies, adversaries and neutrals.

Enable Informed, Decisive Action: Navy commanders must maintain sufficient decision space to operate within an adversary's decision cycle and must continually decide upon available warfighting options as the environment evolves.

To enable informed, decisive action within A2/AD environments, the Navy will be required to enhance operational and tactical decision support to increase warfighting options.

Current State/Near-Term Capabilities (2013-2019):

A number of changes are programmed to improve the Navy's ability to manage, collect, process and disseminate essential information to Navy units afloat and ashore. Near-term BA capabilities to be addressed during this timeframe are as follows:

- Centralized fleet collection management (CM) and tasking to support both standing and ad hoc requirements across diverse intelligence disciplines will be developed and will include: integrated collection plans, a consolidated collection requirements database, and the means to visualize available collection assets;
- Manned, unmanned and other diverse platforms will be fielded to extend organic sensor capability and capacity; commercial space-based imaging and Automatic Information Systems (AIS) collection systems will proliferate; and an emergent capability to task, plan and direct organic and other sensors will be developed;
- The transformation to distributed network environments will begin to emerge; high-performance computing will better understand and predict the physical and virtual environments; automation of multiple intelligence source (multi-INT) data/information fusion and correlation will evolve and adopt common data models and standards; and service-based architectural frameworks will be developed to enhance information sharing;
- An emerging system-of-systems approach for providing a Common Operational Picture (COP) and a Common Maritime Picture (CMP) will begin to shape tactics, techniques and procedures to enhance multi-domain awareness, and enterprise services and visualization tools will be developed to help understand information, depict actions and trends in near real-time;
- Data sharing will be enhanced, enterprise solutions will be pursued for data purchasing and cross-domain solutions will be developed to begin consolidating Top Secret and Secret networks into a single classified domain;
- MOCs will remain as the centers of gravity for Fleet BA coordination and will serve as central nodes for the C2 of Navy ISR assets and resultant in-theater analysis, and fleet commanders and TAOs will be able to establish and share a maritime COP that is periodically updated;
- Planning tools with Theater Security Cooperation capabilities and mission partner capability information sources will become better integrated;
- Improved BA training for all operators and watchstanders will be developed.

Advanced Capabilities (2020-2028):

The vision for Navy's future BA capabilities is being driven by anticipated operations in what is expected to be a more heavily contested and hostile C2 and operational environment. To optimize BA functions in such environments, the Navy will require: 1) improved strategic/operational/tactical sensor coordination and collaboration; 2) integrated intelligence and operational information; 3) better data and information access and sharing; and, 4) advanced information fusion, analysis, dissemination, management and relay. The future BA vision is characterized as follows:

ATLANTIC OCEAN (Feb. 9, 2013) An AV-8B Harrier lands on the flight deck of USS Kearsarge (LHD 3).

Fuse Essential Combat Information:

1) *Streamline Tasking, Planning & Direction:* Navy's BA capabilities require the automation of requirements and task management, the integration of collection planning, more flexible and plentiful ISR capabilities, and accelerated testing and evaluation. Such advanced capabilities involve:

- Fully integrate collection plans, build a single database to manage Navy-wide collection requirements, and develop a display to visualize all available collection assets (to include non-traditional partners); shorten time to respond to commander's Priority Information Requirement (PIRs) and Commander's Critical Information Requirements (CCIR) by automating tasking and collection management processes; enable unmanned sensors to task themselves intelligently based on pre-planned direction and display operational plans and any impacts that might affect the plan;
- Develop and improve collection and integration capabilities of full motion video (FMV) with national imagery systems to conduct precision targeting;
- Provide cross-domain, cross-Service dynamic tasking/re-tasking of assets that permits multi-dimensional, multi-INT correlation and fusion. This re-tasking must also support a Tasking, Collection, Processing, Exploitation and Dissemination (TCPED) process which can dynamically respond to maritime component and COCOM requirements, and changing environments;
- Fully automate ISR CM capabilities to automatically task ISR assets and sensors in response to fleet requirements; employ artificial intelligence to detect gaps in the User-Defined Operational Picture (UDOP) and automatically task ISR assets to fill in information gaps.

2) *Advance Sensor Development Across All Domains:* Navy's increased capability and capacity should be aimed at having complete visibility into regions of interest and overcoming an adversary's efforts to deny access to critical areas. Advanced technologies are needed to:

- Increase autonomy of sensors, platforms, and data processing to reduce data latency and manpower costs;

- Optimize the mix of autonomous manned and unmanned platforms and sensors to increase capability and capacity to commanders and weapons;
- Develop sensors that:
  o Enhance multi-INT, multi-domain collection capabilities, and improve visibility into contested battlespace;
  o Increase the number of multi-purpose, low-cost, networked, deployable, and expendable assets;
  o Process data locally and disseminate required data smartly;
- Replace larger single-capability satellites by hosting sensor payloads on other types of platforms and identifying other options for necessary space-based collections:
  o Investigate micro-, mini-, and nano-satellites to provide fine-scale temporal and spatial resolution data via common C2 paths as well as to provide persistence, increase collection fidelity and mitigate capacity issues;
  o Improve low-power sensors and battery and fuel cell technology to enhance persistence and endurance;
  o Develop and collect strategic signals of interest by improving NTM and tactical collection capabilities in support of emerging threats;
  o Pursue satellites technologies with greater on-board processing and direct downlink capabilities to deployed forces and weapon systems to reduce the SATCOM requirement.
  o Determine the capability and availability of other agencies' and countries' ISR assets in near real-time to support Navy/maritime collection requirements, and make greater use of international partnerships; achieve net-enabled cognitive interactions between disparate forces to enhance collaborative operations, including allies;
- Develop sensors and networks that use "Spectrum Agility" to work seamlessly across broad areas of the spectrum, increasing survivability and effectiveness; detect and precisely measure and map the EM environment in real-time;
- Meet the growing data demand coming from new SIGINT and ocean-based sensors, as well as higher resolution persistent sensors (including FMV) coming from space-based systems and multi-spectral sensors.

3) Fully Automate Processing, Fusion & Product Delivery: The networking, data access and data sharing capabilities described in Chapter One are critical foundations for improved BA. Navy's BA capabilities will require the advent of pervasive cloud-computing technologies, the adoption of common data standards, and enhancements in multi-INT fusion. Dramatically increased computing capability should allow for the real-time transport of vast quantities of data. Advanced technologies are needed to:
- Establish coherent data strategy policies and associated technologies to automatically fuse disparate type of data, which would deliver significant improvements in the ingestion, tagging, indexing, storage, access, and backing-up of essential data;
- Individual platforms would have the ability to manipulate and adjust emissions to maximize operational capability and reduce vulnerability to detection or jamming;
- Shift more data-intensive computing functions to the cloud environment, where analytics of massive data may be conducted:

- Develop a range of software applications, advanced analytical tools, widgets, and decision aids, powered by advances in artificial intelligence (AI), that would move routine functions away from the sailor;

Understand the Operating Environment:

1) Develop a Shared, Relevant Real-time COP/CMP: A shared, relevant real-time COP/CMP will require a more advanced data centric strategy (i.e., architecture) that allows rapid extraction of operational and intelligence data whenever needed. Advanced technologies are needed to:

- Develop and field visualization tools to depict actions and trends in tactical, operational and strategic engagement, and at all phases of engagement; refine the means for assembling disparate information into actionable knowledge;
- Evolve the CMP to an enhanced UDOP with shared situational awareness of all domains—sea, air, land, space, and cyberspace as well as EM spectrum useable by others; gain access to a wider array of data resources and systems, tailorable to individual warfighter's needs, and feed a single, commonly-accessible UDOP available to other users;
  - o Update the COP with operationally-relevant information of friendly and other service forces;
  - o Develop combat systems that automatically report their operational status via their inherent architectures;
  - o Display a spectrum landscape that is mapped using GPS coordinates and is placed into a database for on-demand extraction;
  - o Provide a Blue force NTDS type display that shares spectrum operational views between TAOs and Fleet commanders;
  - o Develop common sensor and track database access and availability that enables user-specific red COP;
  - o Develop new analytical tools that maximize each individual user's ability to customize environmental and operational data for the task at hand.

2) Understand and Predict the Physical and Virtual Environments: Navy requires improved insight into an adversary's actions, intentions, and locations. Navy's future BA capabilities will require improvements in EM spectrum and weather analysis and prediction. Advanced technologies are needed to:

- Understand the complex EM environment to the point that we can assess and predicatively manipulate it to our advantage;
- Employ cloud computing and advanced online analytical tools and databases to allow afloat and ashore users uninterrupted access and enhanced collaboration regardless of location;
- Employ enhanced tools to better visualize targeting information, Combat Identification (CID) reports, Collateral Damage Estimate (CDE), Battle Damage Assessment (BDA), and applicable Rules of Engagement (ROE).
- Advance METOC processing and analysis capabilities to use distributed, integrated computing nodes for cloud operations, and run models with the best numerical characterizations tuned for optimal performance over expanded time scales throughout all domains;
- Access predictive weather, spectrum and network models that are fully coupled with land, air, ocean, ice, and cyber models and incorporate

ensembles, variable resolutions, and confidence levels to produce more accurate 10-day weather forecasts;
- Communicate with fleet or Joint systems to enable exploitation of the physical environment through the next generation of decision planning tools.

3) Understand Capabilities and Intentions of Allies, Adversaries and Neutrals: Navy should provide warfighters with an extensive reach-back capability. Navy's BA capabilities require improvements in automated processing, fusion and product delivery systems. Advanced technologies are needed to:
- Project relevant or suspicious activities of interest onto a variety of prioritized display platforms and automatically alert users to issues, activities, events, or incoming reports as defined by the user.
  o Move from reacting to anticipating adversary behavior and develop means to improve predicting behaviors;
  o Incorporate contextual information from non-traditional sources;
- Provide accurate, real-time battle and Electromagnetic Battle Management (EMBM) assessments and status of own forces in the face of growing threats;
- Develop advanced tools to enable automatic and reliable ingestion of combat ID information from a variety of combat systems and data links;
- Employ robust data replication for situational awareness, plans, and effects assessment across command centers, echelons, and the Services, configurable and tailorable to meet the needs of users at each operational and tactical level;

Enable Informed, Decisive Action:

1) Increase Warfighting Options: Navy's BA capabilities require the means to integrate decision tools to enable improvements in information sharing among warfighters at the tactical and operational levels through discovery and sensor data access to IC or Joint clouds in real time. Advanced information technologies are needed to:
- Enable MOCs to maintain situational awareness across multiple data links simultaneously;
- Improve training simulation to provide a sustained informational edge over potential adversaries. Fleet synthetic trainers should be able to use a fully emulated 4-D physical environment that provides the utmost realism for effective training, modeling and simulation.

ATLANTIC OCEAN
(Nov. 2, 2012) Flight deck
operations aboard
USS Wasp (LHD-1).

PACIFIC OCEAN (Feb. 20, 2012) SEALs and divers from SDVT 1 swim back to USS Michigan (SSGN 727) during an exercise for certification on SEAL delivery vehicle operations.

Chapter 3

Integrated Fires

SUSTAIN THE INITIATIVE, DISRUPT ENEMY INTENTIONS

Overview: Navy Integrated Fires (IF) coordinates all elements within the blue kill chain and disrupts red kill chains in order to seize and hold the initiative in combat, and to limit an enemy's freedom of maneuver and action. The Information Dominance capabilities within Navy IF are the culmination of the Assured C2 and BA functions described previously, which enable the delivery of essential and timely combat information to Navy commanders, deployed units, and weapon systems. Navy IF capabilities are primarily being pursued to: 1) coordinate and synchronize the use of both kinetic and non-kinetic capabilities to achieve desired lethal and non-lethal effects; 2) support all missions and target sets; 3) be applicable in and across all domains (sea, air, land, space and cyberspace); and, 4) be effective across all warfare environments, to include A2/AD scenarios. The growing capabilities in non-kinetic fires expands options for Navy commanders, and is useful in situations where it may be more effective, when kinetic collateral damage risks are deemed unacceptable, where kinetic inventories are limited, or when non-attribution attacks may create an asymmetric strategic, operational or tactical advantage. Major elements of Navy IF and corresponding Navy concerns in each of these areas are highlighted below:

- Employing the full EM Spectrum to Enable Kinetic Fires: Effective utilization of the EM spectrum can optimize the coordination and targeting of conventional "iron-on-target" fires to maximize the warfighting capabilities of Navy and Joint platforms and weapons. *Navy's current capabilities in this area are constrained by the rapid worldwide growth of sophisticated weapons, EW systems, intelligence, operations, and communications systems that greatly complicate the Navy's usage of the spectrum, and by the increasing commercial and legislative demands for reallocating portions of the EM spectrum.*

- Using the Spectrum to Enable Non-kinetic Fires: Non-kinetic fires leverage the EM spectrum as a weapon against an adversary. Non-kinetic fires include offensive cyberspace operations, jamming and the use of directed energy weapons, any of which can be just as destructive as kinetic fires. *Navy's current capabilities in this area are likewise constrained by increasing EM demands, the need for tools to better understand the EM environment, and the limited understanding and application of cyber authorities and effects.*
- Coordinating Kinetic and Non-Kinetic Fires to Achieve Desired Effects: The coordination and synchronization of kinetic and non-kinetic fires across multiple domains offers the advantage of being able to overwhelm an adversary to achieve desired effects. *Navy's current capabilities in this area are constrained by cultural and security barriers and the extensive coordination requirements to minimize unintended effects.*

Functional Descriptions: Two primary warfighting functions associated with Navy IF capabilities will be addressed in the remainder of this chapter, and are briefly outlined below:

Disrupt/Deny/Defeat Red Fires: Naval commanders must always be prepared to defend the force against offensive fires. The threat posed by red forces can be mitigated by pro-active measures that anticipate and counter red, left-side-of-the-kill-chain actions long before these forces can pose a threat to Fleet units. This pre-emptive approach hinges on dissecting red kill chains and disrupting, denying or defeating critical links within those chains, especially in high-threat A2/AD warfighting scenarios. This future capability requires the Navy to prevent an adversary from initiating kinetic and non-kinetic operations by disrupting adversary C2 and preventing effective targeting by kinetic and non-kinetic weapons. This will have the added benefit of reducing reliance on blue high-risk conventional operations.

Enhance Blue Fires: In light of ever-increasing improvements in the speed, accuracy, range and lethality of maritime-related weapons being developed worldwide, Navy commanders must continually seek new means to maximize their own warfighting effectiveness and enhance their ability to deliver accurate and timely fires in all domains. Such an imperative is particularly applicable to future A2/AD warfighting scenarios. This capability requires the Navy to dynamically collaborate across all missions, domains and with the other Services in order to exploit the spectrum as a weapon and integrate targeting and fire control capabilities to enable increased weapon range, effectiveness and lethality.

Current State/Near-Term Capabilities (2013-2019): Over the next six years, a number of changes and improvements are programmed that will enhance the Navy's current ability to integrate fires across the Fleet. Evolving A2/AD capabilities by potential adversaries are driving the need to orient Navy's warfighting capability towards increased integration and interoperability of platforms, sensors, weapons, and systems in line with efforts such as the Air-Sea Battle (ASB) concept. These initiatives are largely being built around the three emerging capabilities and functional areas of Naval Integrated Fire Control – Counter Air (NIFC-CA), Offensive Anti-Surface Warfare (OASuW), and Electromagnetic Spectrum Operations (EMSO). Highlights programmed over the 2012–2019 timeframe involve the following:

- Fielding initial assured C2 capabilities that mitigate existing tactical data link capability and capacity shortfalls in A2/AD environments, such as Link 16 Concurrent Multi-Netting Four Channel (CMN-4) and netting Line of Sight (LOS) sensors through Tactical Targeting Network Technology (TTNT);
- Defining and fielding initial targeting and fire control capabilities that contribute to integrating fires, such as NIFC-CA From the Air (FTA) and From the Sea (FTS);
- Fielding initial increments of Net Enabled Weapons (NEW) such as the Joint Stand Off Weapon (JSOW-C1), Small Diameter Bomb (SDB)-II and OASuW weapons;
- Development and fielding of counter-red C4ISR decoy and deception capabilities (including space-based and traditional systems) to thwart adversary knowledge of Blue forces and integrating these into our combat systems;
- Evolving EA capabilities to meet emerging threats and refining the concept of EM Attack as a broader, coordinated EA capability.
- Examining opportunities and developing solutions to leverage Joint and National capabilities to integrate fires and include them in Concept of Operations (CONOPS) and Concept of Employments (CONEMP) development activities;
- Researching and developing directed energy weapons and their expected CONOPS;
- Continuing refinement of emission control (EMCON) operational concepts;
- Defining and executing experimentation campaigns that inform the full Doctrine, Organization, Training, Material, Logistics, Personnel, Facilities, and Policy spectrum for integrating fires;
- Conducting Modeling, Simulation and Analysis from the engineering to the campaign level in order to quantify potential warfighting returns of various IF capability proposals;
- Refining existing or developing new requirements for integrated fires capabilities.

Advanced Capabilities (2020-2028): Closing current and anticipated capability gaps associated with the Navy IF functions described above requires a coordinated effort across multiple Navy and Joint organizations to consolidate Navy IF initiatives into the first increment of a Navy Integrated Fires Family of Systems (FoS) approach. Additional new fires capabilities such as Directed Energy Weapons (DEW) and offensive cyberspace operations must be incorporated into conventional operations to enable more focused lethal and non-lethal effects. Other Navy IF enhancements will require disparate Navy systems, including C4ISR and combat sensors, targeting, fire control, unmanned systems, and certain weapons, to be network-enabled, which means that platforms, sensors, weapons, and systems will be able to seamlessly exchange combat data and information in an automated real-time manner. This execution network will freely exchange data through, and be an extension of, the Dynamic Flexible Grid (described in Chapter One). Emerging threats in the A2/AD environments will be the major drivers for extending future integrated fires functionality in all missions, and will represent a significant Information Dominance contribution to the ASB concept. Ultimately, Navy's goal will be a

FoS that blends all kinetic and non-kinetic fires into a unified, coherent capability to include unmanned systems, NEWs, DEWs, and cyberspace weapons. Such a FoS would deliver timely and capable effects and prevent the adversary from initiating kinetic or non-kinetic operations, and reduces the reliance upon, or in some cases, the need for high-risk conventional operations. This capability will be characterized by decentralized, inter-domain information sharing environment and feature dynamic, distributed, synchronized and robust cloud infrastructure at the tactical edge that enables real time coordination and integrated response between sea, air and cyberspace forces. This capability will enable Joint Force commanders to employ an array of capabilities to defeat A2/AD threats. Future Navy IF goals to achieve these aspirations are highlighted below.

Disrupt/Deny/Defeat Red Fires:

1) Prevent the adversary from initiating kinetic and non-kinetic operations: Future Navy IF capabilities should provide the COCOM/Joint Task Force Commander (JTF) with a baseline ability to counter an evolving generation of red force C4ISR & targeting in order to maintain U.S. freedom of action and deter aggression. Advanced technologies are needed to:
- Through comprehensive indications and warning, identify "prior to launch" blue fires that could prevent adversary actions against blue forces, to include fielding new capabilities that thwart adversary knowledge of Blue forces, leveraging Joint and National capabilities and fully integrating capabilities from the campaign level down to tactical operations. Efforts would include:
  o Preventing an adversary from obtaining/maintaining identification and targeting and ascertaining intent of blue forces;
  o Preventing an adversary from effective C2 of platforms, weapons and sensors that are required to initiate offensive operations;
  o Integrating cooperative EW systems as part of an EM Spectrum Operations capability that synchronizes strategic, operational, and tactical operations;
  o Integrating manned and unmanned systems to cooperatively work together to counter adversary ISR&T;
  o Enhancing usage of passive kill chains to reduce detection of blue forces;
  o Deceive the adversary by leveraging electronic decoys and stealth/low observables that involve active and passive electronic spoofing of adversary systems, long duration, active electronic off board decoy system (vehicle, payload, and ship controller) and an onboard/off board soft kill coordinator;
- Employ cyber-based military deception (MILDEC) tactics as needed to slow or confuse adversary planning and targeting efforts;
- Develop a cyber exploitation and attack capability to predict and defeat cyber attacks before they occur, to include preventing adversary weapons launch by defeating "weapons on the rail" or in the silo;
- Fully integrate an offensive cyber capability into military operations which would:
  o Scale from preparing the battlespace in major campaigns to locating and identifying high value targets in Irregular Warfare;
  o Include the ability to target adversary infrastructure such as electric power, transportation systems, and other essential needs to high level government

and military functions down to the C2 and tactical systems level;

o Maximize effects to the extent that the adversary will be incapable of bringing kinetic and non-kinetic assets to bear;

o Include the ability to impact adversary weapons/effects prior to or during engagement (such as the ability to take over an adversary weapon's mission system or re-target it against the adversary);

o Combine offensive cyber with RF decoys, spoofing, low observables, and passive kill chains to force the adversary into active sensing modes, allowing U.S. forces to more rapidly target and neutralize targets;

• Develop means to rapidly undertake operational contingencies such as EMCON that would inhibit an adversary's ability to predict the ingress and movement of blue forces.

2) Prevent effective employment of adversary kinetic and non-kinetic weapons: Future Navy IF capabilities should provide COCOM/JTF with a baseline ability to counter an evolving generation of cruise, ballistic, air-to-air and surface-to-air missiles and electronic warfare and offensive cyber weapons in order to maintain U.S. freedom of action and ability to deter aggression. Advanced technologies are needed to:

• Improve IF against adversary conventional and unconventional weapons during all phases of engagement to prevent successful weapons employment to include:

o Manipulating the EM spectrum to defeat adversary weapons during engagement;

o Using hard kill, DEWs and EM-driven weapons against adversary launch platforms and weapons;

• Improve IF against C4ISR&T operations that directly support weapons engagement, to include:

o Preventing C4ISR&T actions to refine targeting;

o Preventing in-flight target updates from reaching weapons;

• Conduct IF through adversary networks or kinetically against critical nodes to actively defeat adversary cyber attacks;

• Field obscurants that are effective in different regions of the EM spectrum, to include:

o Physical obscurants that can be deployed to degrade performance of adversary sensors through reflection, refraction, and/or absorption.

o Deployment of obscurants to adversely impact tactical, operational, and strategic capabilities of the adversary;

ARABIAN SEA (Dec. 13, 2012) USS Jason Dunham (DDG 109) fires its MK 45 5-inch lightweight gun during a live-fire exercise.

- Create the means to identify, secure and use EM spectrum corridors during operations in hostile A2/AD environments, to allow continued freedom of maneuver by friendly forces;
- Maximize low observable operations that leverages low signature forces, passive kill chains, and national assets in support of time and accuracy sensitive tactical functions (such as targeting and ID) and RF deception capabilities.

3) Reduce Blue force reliance on High-risk Conventional Operations: A key outcome of a fully Integrated Fires capability would be the increased warfighting options and improved effectiveness available to commanders through all phases of operations. Navy's IF capabilities must be capable of disabling adversary kinetic and non-kinetic capabilities prior to hostilities. Advanced technologies are needed to:

- Conduct offensive cyber operations and deny adversary spectrum in order to minimize and/or negate the need for conventional kinetic operations;
- Synchronize tactical cyber actions with Fleet operations to increase overall effectiveness, enhance lethality, and reduce operational risk;
- Increase the geographical range, precision and spectrum provided by the Integrated Fires FoS capabilities to increase options and to operate from sanctuaries;
- Increase the role of unmanned systems to improve effectiveness of ISR&T, provide additional platforms and options for kinetic and non-kinetic attack, and reduce risks to warfighters;
- Increase precision of targeting and fire control to prevent collateral damage;
- Increase the effectiveness of our deception efforts to prevent the adversary from using the EM spectrum for targeting and engagement.

Enhance Blue Fires:

1) Integrate Targeting and Fire Control Capabilities: A comprehensive Integrated Fires capability would efficiently use all available sensor data, even from traditional ISR or combat systems, to develop targeting and/or fire control solutions for any weapon and operate completely inside of the adversary's strategic, operational, and tactical decision cycles. This will enable forces to defeat adversary C2 nodes, search and targeting radars, and weapon systems through a combination of defensive maneuvers, enhanced EM countermeasures, advanced long-range kinetic weapons, and cyberspace capabilities in a synchronized fashion. Advanced technologies are needed to:

- Coordinate kinetic and non-kinetic targeting and fire control capabilities to maximize lethal and non-lethal effects and optimize capacity of kinetic assets;
- Deliver dynamic, integrated fire control capability across all missions and domains, to include use of an Integrated Fires FoS to expand sensors beyond traditional ISR to achieve flexible, real-time, targeting and fire control quality data.
- Develop autonomous weapons that are less reliant on their launching platform, able to utilize any sensor in the Joint Force, and that can deliver kinetic and non-kinetic effects
- Develop network-enabled weapons that can share sensor data with other weapons and platforms as well as to act as penetrating sensors for enhancing targeting and fire control;
- Develop networks with the capability and capacity to integrate effects chains across multiple domains and throughout the A2/AD environment to enable C2 of engagements and more efficient use of weapons;
- Fully integrate unmanned systems and net-enabled weapons into strategic, operational, and tactical operations;
- Enable real-time, seamless C2 of cyber operations integrated with kinetic operations.
- Enable flexible active and passive sensor coordination to provide targeting and mitigate capacity and capability shortfalls;
- Expand the means to collect greater fidelity targeting information supporting all missions by correlating multi-INT data and information, including radar systems, passive sensing systems; and video/imagery systems;
- Develop an improved capability to rapidly conduct target systems analysis supporting both kinetic and non-kinetic actions, Collateral Damage Estimation (CDE) and precise aim point mensuration;
- Ensure Integrated Fires FoS supports automated theater-wide CID with the ability to share raw ID data and intelligence and provide confidence levels to the Warfighter;
- Field Automated Battle Management Aids (ABMA) and Dynamic Weapons Coordination (DWC) across all domains to assist the commander in choosing the right weapons for the right targets;
- Integrate long-range directed energy and EM-driven weapons as an integral part of the IF FoS;
- Develop cross-domain tactical targeting applications for Integrated Fires FoS that enable use of organic sensor data to support operations without need for external support as well as network-enabled geo-location mission applications that correlate and fuse multi-INT sources for targeting.
- Develop artificial intelligence algorithms to support automation and collaboration;
- Develop collaborative functionality interoperable from the RF signal in space to the data and information being processed at the machine level across platforms and domains.

2) *Exploit the Spectrum as a Weapon:* In the future, Navy's IF capabilities will enable forces to promptly act across domains to neutralize A2/AD threats by degrading, disrupting or destroying adversary space, air, maritime and land-

based sensor and communication links through a myriad of electronic and cyber resources. Advanced technologies are needed to:

- Advance EM Maneuver Warfare concept as a primary element of warfighting;
- Monitor the complex EM environment to the point that the natural state of the EM spectrum can be understood by:
  o Assessing and predicatively manipulate it to our advantage to support integrated fires;
  o Masking blue forces and enhance detection of red forces;
  o Manipulating the spectrum as a fires capability, including directed energy weapons (DEWs) such as high-power microwave, lasers, and RF systems;
- Develop and refine Offensive Cyber Operations that can impact single, isolated computer and combat systems to global wired or wireless networks;
- Incorporate DEWs and cyber as a non-kinetic fires capability that can deliver lethal effects;
- Leverage all available areas of the EM spectrum to ensure the capability to provide targeting data for blue forces by:
  o Developing the capability to baseline the natural state of the EM environment prior to military operations;
  o Developing and enhancing existing systems that detect and measure perturbations in the natural environment;
  o Supporting correlation and fusion to enhance overall information pedigree for targeting and other critical functions.
- Modernize current MILDEC tactics, techniques and procedures to take advantage of the full range of EM capabilities.

3) Enable increased Weapon Range, Effectiveness & Lethality: Navy's IF capabilities must be capable of maximizing the full capabilities of all the weapons at our disposal, both kinetic and non-kinetic, in any environment or domain. This includes capability in sophisticated adversary A2/AD environments. Advanced technologies are needed to:

- Improve blue C4ISR capabilities to support long-range fires, allow direct weapon-to-weapon coordination, and provide sufficient fidelity;
- Develop ABMA/DWC capabilities to optimize target/weapon pairings;
- Develop netted sensors and long range theater wide CID that can provide persistence and allow employment of longer range weapons, to include:
  o Expanding the suite of sensors to increase availability of critical targeting nodes over a greater length of time, including using weapons to share sensor data to enhance lethality;
  o Increasing the accuracy of targeting data and information through multi-INT correlation and fusion.
- Ensure the Integrated Fires FoS provides persistence, delivers greater fidelity, and permits development of long range weapons to mitigate A2/AD risks, to include:
  o Using unmanned systems as shooters and NEWs to support tactical operations;
  o Ensuring information is not lost, altered or loses fidelity to further support operations in the A2/AD environment;

4) Dynamically Collaborate across all Domains and Services: To achieve persistent, credible combat power to more effectively attack adversary targets concurrently in multiple domains and throughout the operational area, Navy's IF capabilities will rely on the ability to transparently share information across the Joint force and in all domains despite sophisticated adversary attempts to disrupt and deny that data flow. Advanced technologies are needed to:

- Ensure all platforms are capable of rapid processing and sharing of time critical data and information, including distributed, synchronized and agile information sharing functionality for the commander and operator at the tactical edge, even in a sophisticated A2AD operational environment;
- Improve information infrastructure and assured C2 collaborative systems that support dynamic, integrated fires (described in Chapter One);
- Develop means to achieve data and mission systems interoperability, to include:
  o Defining common operating environments and data interoperability standards (i.e., normalized data) across platforms, sensors, weapons, and systems;
  o Delivering organic PNT functionality across the Integrated Fires FoS;
  o Facilitating an application environment enabling machine-machine transactions;
- Work collectively to protect the Blue EM Spectrum;
- Develop common architectures and common processes for planning and execution across domains and services;
- Enable net-enabled cognitive interactions between disparate forces to enhance collaborative operations;
- Ensure seamless command and control capabilities across domains for all fires;
- Develop robust, secure, and safe capability that enables unmanned platforms and weapons capable of "fire and forget" functionality to autonomously target and engage hostile forces without higher echelon tasking;
- Field advanced tactical data links and tactical network systems to alleviate capability and capacity issues of ISR assets and network systems in an A2/AD environment;
- Refine the means to rapidly assemble disparate information to enable a commander's actions;
- Incorporate new unmanned systems into the Integrated Fires network to support integrated fires with sensor data, tactical network connectivity and data processing;
- Synchronize kinetic and non-kinetic fires with tactical and operational level battle management applications resident on the network;
- Provide cross-domain, dynamic tasking/re-tasking of assets that permits Multi-dimensional optimization (i.e., platform and domain transparency) to increase efficiency across the kill chain by improving weapon/target pairing;
- Institute a dynamic TCPED process and flexible ad-hoc architecture to maximize sensor capability and capacity.

GULF OF THAILAND (Feb. 15, 2013) A naval aircrewman surveys the Gulf of Thailand from an MH-60S Sea Hawk helicopter assigned to the HSC-25, embarked with the forward-deployed amphibious assault ship USS Bonhomme Richard (LHD 6).

SUMMARY

A2/AD weapons, systems and approaches to warfare are being developed by a number of our adversaries to disrupt or negate the Navy's current technological advantages and preferred modes of warfare. This roadmap depicts Navy's required future Information Dominance capabilities based on the anticipated operating and information environments expected during the 2013–2028 timeframe. This document is intended to help synchronize and integrate Navy's diverse information-related programs, systems, functions and initiatives to meet future warfighting objectives. Required improvements in these areas will involve the Navy's intellectual, technological, human and financial resources.

Assured Command and Control: Assured C2 is ultimately about ensuring a commander's ability to command assigned forces to achieve the tactical, operational or strategic objectives established by the chain of command. Navy's future information infrastructure must be able to maintain essential network and data link services across secured segments of the EM spectrum in high-threat scenarios to transport, share, store, protect and disseminate critical data and combat information required by forward deployed units and on-scene commanders. Of particular importance is the "operationalization" of the EM spectrum and cyberspace into a warfighting domain, to enable use of the EM spectrum as maneuver space.

Battlespace Awareness: The Navy will face an increasingly complex and dynamic strategic environment over the next fifteen years. Threats imposed by both stateless actors and traditional states demand that we take an active, layered defense-in-depth approach. To defeat such threats will require vastly improving and structuring Navy's overall information content for use in combat, and will require improved means to rapidly sense, collect, process, analyze, evaluate and exploit critical intelligence regarding our adversaries and the operating environment. Our future information content will serve as the basis from which nearly all timely decisions will be made during information-intensive combat, enabling our forces to more effectively maneuver the force and coordinate actions in order to target and engage adversaries inside their decision cycles.

Integrate Fires: Navy Integrated Fires will focus on coordinating all elements within the kill chain in order to seize and hold the initiative in combat, and to limit an enemy's ability to maneuver and act. By coordinating and synchronizing the use of all available Navy and Joint kinetic and non-kinetic capabilities, the Navy will be

optimally positioned to achieve desired lethal and non-lethal effects across the full array of warfare environments, to include A2AD scenarios. Integrated Fires will require new capabilities to fully employ integrated information effects in warfare by expanding the use of advanced electronic warfare and offensive cyber effects to complement existing and planned air, surface and subsurface kinetic effects within the battlespace. Future information effects will be designed to impact and change adversary behaviors, or control, manipulate, deny, degrade or destroy their warfighting capabilities.

ANDAMAN SEA (Oct. 12, 2012) USS John C. Stennis (CVN 74) and USS George Washington (CVN 73) underway in the Andaman Sea with ships from their Carrier Strike Groups. The U.S. Navy is reliable, flexible, and ready to respond worldwide on, above, and below the sea.